easy internet
and email for
the over 50s

Bob Reeves

Hodder Education
338 Euston Road, London NW1 3BH.

Hodder Education is an Hachette UK company

First published in UK 2011 by Hodder Education.

www.hoddereducation.co.uk

Typeset by MPS Limited, a Macmillan Company.
Printed in Great Britain by CPI Cox & Wyman, Reading.

Contents

1

setting up your email address

An email address is the first thing you need in order to send and receive email. This email address, which will always include the @ symbol, will be unique to you. There are two ways to set up your email: you can use an email software program loaded onto your PC; or you can use a web-based email which you can access from anywhere with an Internet connection. This chapter will guide you through both options.

1.1 Email software

There are two main types of email software. The first is standard **email software** such as Microsoft Outlook, which comes as part of some versions of the Microsoft Office package. As an alternative you can use Windows Live Mail, which is available as a free download. If you do have Outlook this will already be installed and set up on your computer and you run it from your desktop in the same way as other programs. If you have had a new computer set up by your supplier, this may have been set up for you. If not, then Section 1.5 of this chapter will show you how to do it.

The second type is called **web-based** email. With this, you do not need to have email software installed your computer, as you can get access to it using the Internet. Web-based email is generally free. For example, Microsoft Hotmail, Google's Gmail and YahooMail are all free-to-use web-based email services. You may also be given free email addresses from your Internet Service Provider (ISP, the company you get your Internet connection from). For example, BT, AOL and Tiscali all provide free web-based email too.

> All email software has the same basic functions whether you use a standard or web-based one. Whichever one you choose, you will need to know your email address (i.e. if a shop has set one up for you) or create an email address if you are setting up for yourself.

1.2 Email basics

Email stands for electronic mail and the easiest way to think of it is as an electronic letter. You do not have to print it out and send it – instead, it is sent electronically over the Internet. Therefore, you must have Internet access in order to use email.

Like normal mail, emails are sent and received using addresses. All email addresses follow the same format. For example: marjorie.franklin@googlemail.com. The bit before the @ sign is usually used to identify the individual, and the bit after is the name of the **email provider**.

1.3 Choosing which email software to use

Choosing which email provider to use can be quite difficult. There is a lot of debate over which is the 'best'. Much of it depends on your personal preferences and the way in which you want to access your emails. It also depends on what software is already on your computer and whether you are prepared to pay for new software if you don't have it already. Most individual users tend to go for a free web-based email.

Reasons to choose web-based email

* You can access your email from any computer. Therefore, if you don't have your own computer, or are away from home and want to access your emails perhaps from a library or Internet cafe, then this is the choice for you.
* If you do not already have email software on your computer and you don't want to buy or download any, then it is simpler to set up and use a web-based email service.
* It's completely free and you can have several different addresses if you want to.
* You get access to other websites. For example, if you get a Gmail account with Google, you can then use their **chat rooms** and **blogging** websites without having to register again.

Reasons to choose standard email

* You may already have the email software on your computer, and it may already have been set up for you when you bought it. If this is the case, this will save you the time of setting up a web-based email.
* Web-based emails can be cut off if you don't use them for a while. Most web-based emails will be cut off after a month if you have not sent or received an email in that time.

* If your connection to the Internet is not available for any reason, you cannot get at old messages with web-based mail, but you can with standard email.
* It is claimed that it is more secure as your emails are stored on your computer rather than on the Internet. See Chapter 7 for more advice on keeping your information secure.

1.4 Setting up a web-based email address

In this example, we will be using Gmail, which is a free email service provided by Google.

1 Open Internet Explorer, by clicking on the icon in the Taskbar, or by clicking on 'Start' and clicking on it from the list.

2 In the address bar, type www.gmail.com as shown:

3 Press ENTER or you can click the blue arrow to the right of the address bar. The Gmail web page will now be displayed in the main window.

4 Many services are provided free on the Internet, but the websites offering them usually make you register with them first. This involves filling in an **online form**. You will get familiar with this process as you use the Internet more, although it might take a while the first time you do it. Move the mouse over the text that reads 'Create an account'. Notice that the pointer changes to a hand with one finger pointing. This pointer image means that you are on a **hyperlink** or **link**. This will take you to another page.

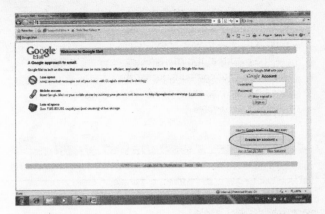

5 Click on the text 'Create an account'. You will now be asked to fill in the form.

6 Click in the box where an answer is needed, e.g. First Name.

7 Type in your answer. Use the SHIFT key if you want letters in capitals.

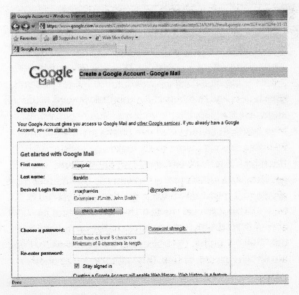

8 Carry on doing this for all of the boxes, filling in the information that is asked for. You can tell which box you are in by looking for the cursor (the small vertical flashing line). If you do anything wrong it will ask you to fill in the box again. In this example, we have set it up for someone called Marjorie Franklin, but you will need to choose the name that you want to use.

Hints and tips

Your email address does not have to be your real name. Remember that you will be giving this address to people so they can email you. Bunnykins@gmail.com might not be appropriate if you are going to be emailing a vicar!

9 After you have typed in your desired email address, click the 'Check availability' button. This will tell you whether someone else has already got that address. If your name is John Smith, then you're in trouble!

10 You are now asked for a password. Passwords mean that only you can log on to this email address, so they are very important. Choose one that you will remember. It also has to be a good password, which means that you have letters and numbers in it. (This makes them harder for people to guess.) Some websites require passwords to have a certain number of characters to make them more secure.

11 You have to type the password again in the box underneath.

12 You need to move down the page to fill the rest of the form in. Use the scroll wheel on your mouse (if it has one) or click several times on the arrow at the bottom of the scroll bar on the right of the screen. This is called **scrolling**.

13 You need to complete the Security Question and Answer box. But you can leave the Secondary email box empty. The idea of the security question is that if you forget your password, you can get a reminder by entering this information.

14 Type the words shown in the Word verification box. This is a security measure to ensure that only humans and not hackers' programs are creating the email account.

15 Scroll down to the bottom of the page and click the 'I accept. Create my account' button. You should see a screen which tells you that you have successfully set up your Gmail account.

16 You will be asked to verify your account, which is sometimes done by sending you a text to your mobile phone. This should be instant and you will be prompted to type in the code that you have received.

Congratulations – you have now set up your email.

1 Click on 'I'm ready – show me my account' and you are ready to send and receive emails.

2 When you want to access your emails all you need to do is open Internet Explorer, go to the www.gmail.com website and type in your user name and password. You can do this on any computer that is connected to the Internet.

1.5 Setting up Microsoft Outlook

If you don't have Microsoft Outlook you can skip this section. The instructions here show you how to set up Microsoft Outlook to use with the Gmail address although it is possible to use other addresses.

In this example we will use Gmail as our email provider, and will set it up so that we can use the Gmail address through Microsoft Outlook.

1 Double click on Microsoft Outlook on the desktop, or go to the 'Start' menu and 'All programs' and click on it from the list. If you can't find it, it probably means that you haven't got it on your computer so you will either have to buy it and install it, or use a web-based email as described earlier.

2 Assuming this is the first time you have used Microsoft Office it will work through a series of screens prompting you for information that you need to set it up. This is called a '**wizard**'.

3 Click on 'Next'.

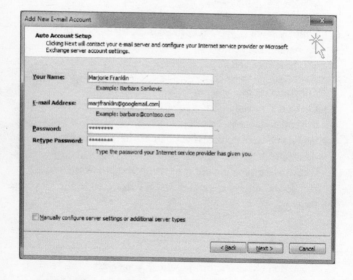

4 Type in the email address you have just set up and the name that you want to appear on your emails when you send them to people, and click 'Next'.

Outlook will now take a few seconds to configure itself and will then open up your email ready to use. The screen will look something like this:

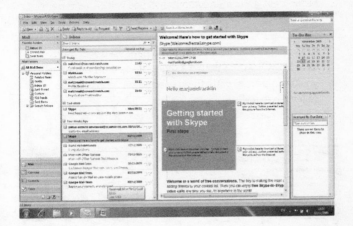

2

sending and receiving emails

If you do not already have an **email address**, you will need to go back to Chapter 1, which shows you how to set one up. As you may have seen, there are lots of email providers to choose from and each one is slightly different. However, they all share some common characteristics.

This chapter will use Gmail (also known as Googlemail) as an example. At the time this book was written, this was one of the most widely used email service. If you are using another type of email provider or program such as Outlook, Yahoo, or Tiscali for example, then your screens will look different to those shown here. However, the basic principles remain the same.

2.1 The basic functions of email

To start with, we will have a look at the standard features of email and what they do.

1 Open Internet Explorer and type www.gmail.com into the address bar.

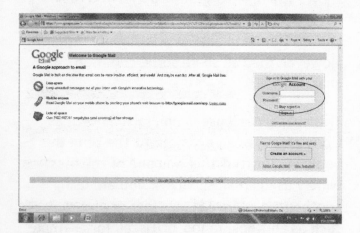

2 Type in your email address and password on the right-hand side.

3 The main email page will now open. Most of what you do can be done from here. We have labelled two of the main areas.

Area A: From here you can create new emails by clicking on the 'Compose Mail' link or you can look through the folders where emails are stored. The main folder you will be using is called your Inbox and this is where any emails you receive will go.

Area B: This part of the page lists all of the emails that are in whichever folder you have clicked on in Area A. The standard setting is for it to display the contents of your Inbox.

Notice that unread messages are usually in bold so that you spot them. Also, next to 'Inbox' in Area A there is a number in brackets which indicated how many emails you have that you haven't yet read.

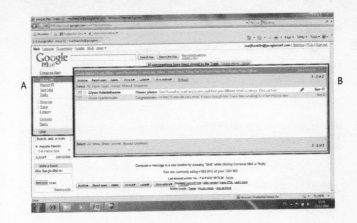

There will already be at least one message in your Inbox from Google Mail welcoming you to the service.

1 To view an email, click on it in Area B. Area B will now change to show the content of the email.

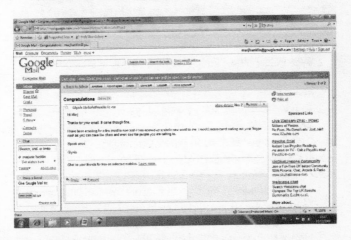

2 Once you have read the message there are a number of things you can do with it such as reply to it, or delete it. We will come back to these later.

2.2 Sending an email

1 Click on 'Compose Mail' in the top left-hand corner. Note that Area B has changed again. This time it is ready for you to send an email.

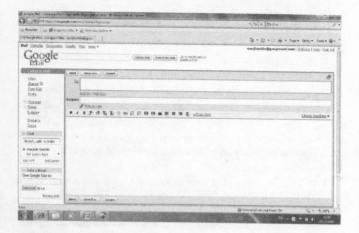

2 You need to know the email address of the person that you want to send this email to. Type the email address into the box next to where it says 'To:'. It is important that you get the address exactly right or your message will not be sent.

3 The 'Add cc' option can be used to send the same email to someone else. If you want to do this, click on 'Add cc' and type the email address of the second person.

4 In the 'Subject:' box, type a heading for what your email is about.

5 Click in the big white box. This is where you will type in your message. Think of this white space as a piece of paper on to which you are writing (typing) the message that you want to send. It can be as short or long as you like.

6 Start your message with a salutation. Common practice seems to be 'Hi', but you don't have to do that. So type 'Hi Glynis' and press ENTER twice. Notice that pressing ENTER starts a new line, so pressing it twice leaves a blank line.

7 You now type your message. You can just type away and it will automatically start on a new line when it needs to. If you want to start a new paragraph just press ENTER twice. Notice that lines have been left between the paragraphs in this case. Also notice the standard use of capital letters, which is achieved by holding down the SHIFT key as you type the letter that you want to be a capital.

8 Common practice is to put your name at the bottom of the email, perhaps with an informal sign off such as 'Regards' or 'Cheers' but there are no hard and fast rules.

9 Click 'Send' in the bottom left-hand corner of this window.

That's it. Your message now flies across cyberspace and will appear in the inbox of the person you sent it to. In most cases, this will be almost instant, but sometimes it may take a few minutes or even hours for the message to be received.

Correcting errors

If you make a mistake when typing use the DELETE key or the BACKSPACE key to delete the error and then re-type.

For example, to change 'Glynes' to 'Glynis' as it should be:

1 Move the mouse to just before or just after the error and click. There is a small line called the **cursor**, which shows you where you are. If you start typing now it will put the text where the cursor is.

2 Delete the letter 'e' by pressing the BACKSPACE key and then type the letter 'i'. Note that BACKSPACE deletes the character to the left of the cursor and the DELETE key deletes the character to the right.

If you want to delete lots of text, you keep pressing BACKSPACE or DELETE. An alternative method is to **highlight** the text you want to delete and delete it all in one go.

For example, to delete the whole of the first sentence:

1 Point the mouse just before the T of 'This' on the first line, then hold down the left button. Move the mouse to just after the exclamation mark. You will see that all of the text in-between is now highlighted.

2 Let go of the left mouse button and press the DELETE key. The whole block of text is gone.

2.3 Saving your email

When you send an email a copy of it will be stored automatically in the 'Sent Items' folder. If you want to go back to an email you have sent previously, you can click on 'Sent Mail', find the email in the list and double click on it to read it.

You might also find it useful to save your email as you go along. For example, if you were typing quite a long email and then something went wrong or you deleted it by accident, you would have to start typing all over again. You can save a 'draft' version of the email so that if this did happen you can go back to the saved draft. To do this:

1 You can save your message at any time when you are typing it by clicking on 'Save Now' from the menu options at the top of Area B. You will see a message telling you that a copy of the email has been saved into the 'Drafts' folder.

2 If you need to get back to the draft version, click on the
'Drafts' folder and then double click on the message.

The message will then be on the screen and you can carry
on typing, or send the message.

2.4 Receiving and replying to emails

When people send you an email it will appear in your Inbox.
You have already seen this in Section 2.1 when you opened the
welcome email from the Gmail team. All the messages received
go into your Inbox and are stored there whether you have read
them or not. When you load your email it will show you how many
new emails you have got. You can then open up your Inbox and read
them. If you are using email for the first time, you might have to
wait for your friends to reply to the one you have just sent, or you
could find someone who you know uses email, give them your new
email address and ask them to email you so you can check it all
works okay.

In this example, Marjorie has got an email back from her friend
Glynis. To read and then reply to an email:

1 Click on the message so that it is **highlighted** as shown.

The message itself will now open.

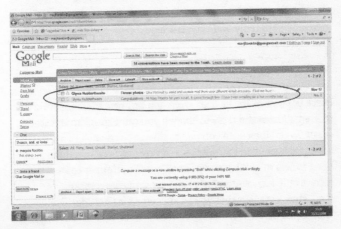

2 To reply to this message, click on 'Reply' from the options.

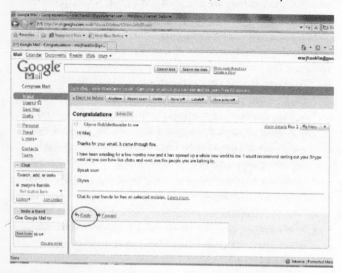

3 A new window will open showing the original message.
4 Replying to a message is almost the same as sending on as described in Section 2.2. It is actually slightly easier because you will see that it has put their email address in automatically so you don't need to remember it.
5 Type your reply and click on 'Send'.

2.5 Printing and deleting emails

Every time you go into Gmail from now on you will need to check your Inbox to see if you have got any new messages. You can then read them and reply to them if you want to. All emails will stay in your Inbox until you delete them. You can delete messages after you have read them, or you can keep them if you think you need to. You might want to print an email out, either to keep as a record, or if you need a paper copy for some other reason.

Some email you get will be junk email (called **spam**) and you should just delete it without even reading it. Some email software has a separate folder called 'Spam' where it automatically puts emails if it thinks they are junk email. As a general rule, you should never open an email unless you recognize who it is from – you should just delete it. Sometimes real emails can get put into the Spam folder by mistake so it is always worth checking to see if you recognize the email address before you delete them.

To delete an email from your Inbox:

1 Click on the Inbox so that all messages are shown. Note that as you receive more messages your Inbox will fill up and you may need to scroll down to see them all. Some email systems will divide the message headers between several pages, displaying only a certain number on each. This can get very confusing after a while so it is a good idea to delete emails that you no longer need.

2 Click on the message that you want to delete.

3 Select 'delete' from the options at the top of the window where the message is displayed.

To print an email:

1 Click on the email you want to print, in the Inbox. It will then open in a new window.

2 Click on the 'Print' icon.

3

sending and receiving email attachments

When you send an email, you are sending some text from one computer to another. If you want to send other things too, you can do this using an **attachment**. Everything that is on your computer whether it is a document, a photograph, a piece of music, or a video clip, is stored in a **file**. In turn these files are stored in **libraries** and **folders**. To send an attachment, you are basically attaching one of these files to your email and it will be sent at the same time as the message.

This does present a small problem in that you need to know which file contains the information you want to send, and which library and folder it is stored in. For example, if you had a photograph that you wanted to send to someone, you would have to know the name of the file that contains the photograph, and the name of the folder that the file is stored in.

3.1 Understanding libraries, folders and files

When you send or receive an attachment you have to send it from, or receive it into a folder. There are lots of folders on your computer already, and you can create your own if you want to. Folders are usually stored in one of four main libraries called Documents, Music, Pictures and Videos. You will add folders to these to store different files, or you can store files straight into the library.

1 To open the 'Documents' library, click on the Start button and select 'Documents'.

The 'Documents' library will be slightly different on every computer, so yours may not look exactly the same as this.

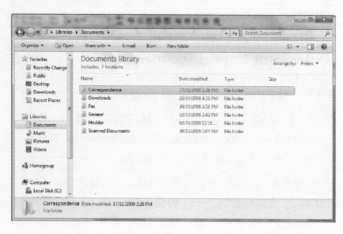

In this example there are some folders shown with the small yellow icon shaped like a folder.

2 To find out what is in a folder, double click on it.

3 To view a different library, e.g. Pictures, click on it on the left-hand list.

In this example, the 'Pictures' library has been clicked on. Inside it there is a folder called 'Sample Pictures' and several photos.

4 To see what is inside a folder, double click on it.

For example, within the Music library there may be an **iTunes** folder containing lots of music files. You will probably be using your computer for other things as well, such as writing letters, or storing digital photographs. If this is the case, you will have come across the libraries, because every time you save anything, you are asked which library you want to save into.

5 Click on the cross to close this down.

We will now look at how you use email to send and receive one of the photos. The same process applies to files that you want to send via email. It would be useful if you could find a file, perhaps a photograph that you can send to someone as a test.

3.2 Sending an attachment

1 Open Google Mail if it is not already open by opening Internet Explorer and typing www.gmail.com into the address bar.
  You will need to log in with your email address and password.
2 Click on 'Compose Mail' in the top left-hand corner.
3 You can now write your email as described in the previous chapter, adding the email address of the person you are writing to, and filling in the subject line, and then typing your message.
4 To add an attachment, click on 'Attach a File'.

Windows Explorer will now open. Yours may not look exactly the same as this.

This is where you need to know the name of the library, folder and the file that you want to attach. If it is a photograph, it is most likely that it will be in the 'Pictures library'. If it is music, it is likely to be in the 'Music library'. Other types of files usually get saved into the 'Documents' library.

5 For this example, double click on 'Libraries' and then double click on 'Pictures'.

6 Click on the file that you want to send. If you want to send more than one, hold the CTRL key down and then click on each one you want to send.

7 Click on 'Open'. In this example we have attached two photos. You can see that they are attached as their file names are shown.

8 Finish the main text of your email and when you are ready, click 'Send' and the photos will be sent with the email.

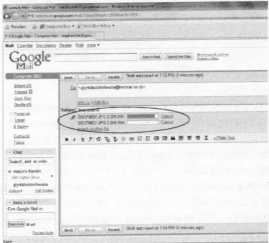

3.3 Receiving an attachment

When you receive an attachment you can view the file and save it onto your computer. For example, if someone sent you a photograph as an attachment, you could view the photograph and then save it onto your computer.

You will know that an email has an attachment, because you will see a paper clip symbol next to the email in your Inbox. In this example, the paper clip is very small but you can see where it will be displayed. The most recent email in the Inbox has an attachment.

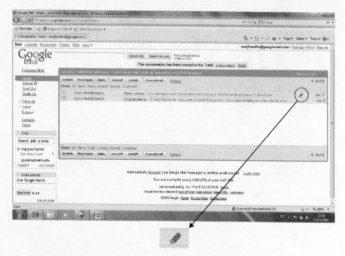

1 Move the mouse to the message and click on it.
2 The message will now open in its own window and you can see the name of the file that has been attached. In this case there are two called 'Flowers 1.jpg' and 'Flowers 2.jpg'. You can see a small version of the image within the email.
3 To open the attachment, click on it. It will then open. As this is a picture, it will open up and let you see the picture.
4 To save the picture onto your computer, close this window and then click on 'Download'.
5 You are prompted to save it into a particular folder. Select the 'Pictures' library on the left-hand side and click on 'Save'.
6 Go back to the email and reply to it, or delete it. You can safely delete the message now because you have saved the attachment onto your own computer. If you deleted the message without saving the attachment first, you would delete the attachment too.

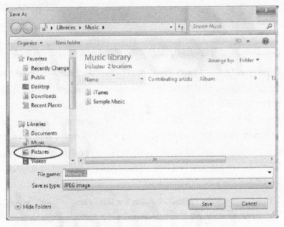

7 You can view the photograph at any time without having to use email. Click on the Start button then on 'Pictures'.

8 Find the 'Flowers1.jpg' file and double click on it. It will then open for you to view.

4

organizing email messages and contacts

Even if you only use email occasionally, it is worth being organized. Every time you send and receive an email, a copy of the email is stored in folders within the email software or service. After a while these folders can get full which can make it hard to find old messages if you need to, and it can make email run slower.

You will also be using email addresses. All email software and services have a facility for storing your contacts' details. This is sometimes called an address book and in that you can store people's names and email addresses.

This chapter will show you how to keep track of the folders and how to manage your **contacts**.

4.1 Organizing the folders

In the last chapter we looked at the way that your computer has **folders** that it uses to store **files**. Email works in a similar way in that it has folders, which it uses to store emails. Let's look at the folders now.

1 Open Google Mail. The folders are listed on the left-hand side.

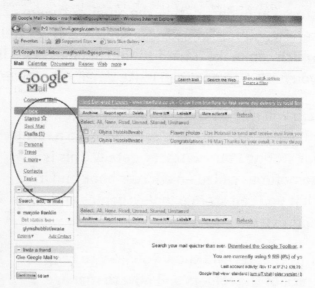

The basic set-up for the folders is that every message you ever receive is stored in a folder called the Inbox. You have already used the Inbox to read messages. The messages will stay there until you delete them. Also, every message you ever send is stored in the folder called Sent Mail. Again, these will stay there until you delete them.

Another useful folder is the Drafts folder. If you save any part-written emails they go in here. Also Google Mail will do an auto-save if you have been typing for a while.

Also notice that there is a link to '6 more' folders. Useful ones in here are the Spam and Trash folders. Spam is where junk emails go automatically but sometimes some genuine ones end up in here.

Trash is where emails go after you have deleted them. Eventually they will disappear from here but you they are still here for a while in case you delete one by accident.

You can access any one of these folders in the same way as the Inbox folder by clicking on it. The contents of the folder are then shown.

Creating folders

* To create a new folder, click on the 'New label' option and type in a suitable name for this folder. This will then be displayed with the other folders in the left-hand side.

Moving messages

All new emails will automatically go into your Inbox. You may want to move messages out of your Inbox into a new folder. To do this:

1 Click in the small box to the left of the message. When you do this a 'tick' will appear. You can tick as many messages as you like here.
2 Click 'Move to' and all of your folders are shown.
3 Click on the folder that you want to move it to, and click 'OK'. The message has now been moved.

You can do other things from this menu using the same method of ticking the messages. One particularly useful feature is that you can tick all the messages that you no longer want and then click on 'Delete' to get rid of lots of un-needed messages all in one go.

4.2 Using the contacts list

You can't possibly remember every single email address that you will ever use and having them all written down somewhere is impractical. That's why there is a contacts facility built in to all email software. The beauty of this is that once you have typed someone's email address into your contacts, you can just click on their details to send them an email – you don't have to type the address in ever again.

1 To open the contacts in Gmail, click on 'Contacts' which is on the left-hand side under the folders.
2 The contacts list will open. It will currently be empty as we have not put any contacts in yet.
3 Click on 'New contact' icon as shown. A screen will be displayed into which you can type all the contact details for one person.

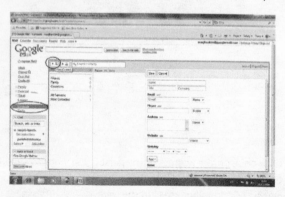

4 You don't have to fill all of this in but you will probably want to type in their name and email address.
5 Click on 'Save' when you have finished.
6 Repeat this process every time you want to add a new contact.

7 To change any of the details, you double click on the person and then click on 'Name' and you get back to the same screen to make changes.

8 Click on the cross to close the contacts.

9 Click on 'Mail' in the bottom left-hand corner to get back to your Inbox.

10 To send an email to someone in your contacts, click on 'Compose Mail' in the usual way.

11 Click on 'To:'. This will open the contacts list.

12 Click once on the address of the person you want to send the email to. Then click on 'Done'.

13 Finish and send your email in the usual way.

In this example there is only one person in the contacts list but you are likely to have lots of people in yours. If you want to send the same message to lots of people, then repeat step 11 for all the people you want to receive the email.

5

getting started on the Internet

The Internet is a worldwide connection of computers. It can be used for communicating and sharing information in many ways, and the most important of these is the World Wide Web (www). This is made up of millions and millions of pages of information, and the links between them. These pages are called **web pages**. A collection of web pages is called a **website**. All sorts of organizations and individuals might create a website. In many cases these are businesses trying to sell things, but there are also websites for government organizations, charities, clubs and private individuals.

This presents a few problems. The first is that there is so much information available that it can be difficult to find what you need. The second is that there is an awful lot of rubbish in among the good stuff. The third is that the information on the Web is constantly changing. New websites are being put up all the time, websites are being removed and the content of individual websites changes.

5.1 Finding a website when you know the web address

The easiest way to find a website is if you know the **address**. Website addresses are unique, so no two websites can have exactly the same name. Most organizations advertise their web addresses and include them in their advertising. For example: www.oxfam.org.uk.

To go to a website if you know the address:

1 Double click on the Internet Explorer icon on the desktop or click on the icon in the Taskbar.

Internet Explorer will now load and a web page will be displayed. The first page to be shown is known as your **home page**. In this case, the home page is Tiscali, which is an ISP. Yours might be something completely different.

Circled at the top of the window is the **address bar**. This is where it shows the address of the page you are on, and where you type the address of the page that you want to go to.

2 Click in the address bar and type in the address of the website you want to go to, in this case www.oxfam.org.uk. It is very important that you type the address exactly as shown with the correct slashes and full stops where relevant.

3 Press ENTER or click on the green arrow to the right of the address bar.

After a few seconds, you will be taken to the page with the address that you have just typed in.

When you get to the page, it is probably the 'home page' of the website. This time 'home page' refers to the main page of a website. It will usually contain general information that welcomes you to the site and tells you about the organization or person who is responsible for it.

Once you are in the website, you may need to move to other parts of the website to find what you want. Nearly all web pages include links to other pages. These links are called **hyperlinks**. They might take you to another page on the same website, or to a page on another website.

Hyperlinks can be attached to anything. There might be a link from a piece of text, or from a picture. They try to make it easy for you to spot the links and explain where the link will take you. The **mouse pointer**, which normally looks like this ⬉ will change when you hover the mouse over an object on a web page. If it changes to a little hand like this 🖑 that means that there is a link to another page.

4 Find a hyperlink (any hyperlink) and click on it. It will probably take you to a different web page, and this may have lots of information on it, and more hyperlinks.

5 You can quite quickly lose track of where you are. After you have clicked on a few hyperlinks, you have lost the page where you started. If this happens and you want to return to it, click the 'Back' button, which is the arrow (pointing left) in the top left-hand corner of the window. This will take you to the previous page.

6 Click the 'Back' button again. This will take you back to the
  page before that, and so on until eventually you are back
  where you started.

Another problem is that sometimes when you click on a
hyperlink, a new window opens up. This means that your original
page is still open in the background. To get back to the original
page in this case:

* Click on the small cross in the top right-hand corner of the
  window. This window then closes, and your original page is
  displayed again.

5.2 Structure of web addresses

It is useful to be able to recognize the way that web addresses
are put together. Sometimes it will give you a hint about the nature
of the site. Most addresses look like this: www.hodder.co.uk.

* The www means World Wide Web and most (but not all)
  addresses start with this. In most cases you don't even
  need to type this in.
* The next part tells you the name of the individual or
  organization who owns the website. In this case it is
  hodder (the publisher of this book).
* The last part of the address tells you what type of
  organization or person owns the website and where it
  is in the world. The table below shows some common
  examples.

.com	Stands for 'commercial' and will be a business. Could be anywhere in the world.
.co.uk	a UK business.
.org.uk	a UK organization, but not a business, e.g. a charity.
.gov.uk	a UK government website.
.ac.uk	a UK college or university. The 'ac' is short for academic.
au, it, de	These are country codes that appear at the end of an address and indicate which country they come from. In this example: Australia, Italy and Germany.

You will notice that once you have typed in an address, Internet Explorer will remember that you have been there. If you go back to this website again, you will only need to type in the first few letters into the address bar and it will list the addresses with these letters in. You can then select it from the list displayed.

5.3 Finding information using a search engine

If you do NOT know the web address, you will need to search the Internet to find the information you need. To do this, you need a **search engine**.

Search engines are free and you can access them using the Internet. There are lots to choose from but the most common ones are Google, Yahoo and Ask. They all do the same thing and it is up to you which one you use. The most popular one at the moment is Google.

A search engine allows you to type in key words that describe what you are looking for. For example, let's say we want to make a donation to the British Red Cross and we need to find the website.

You might start by searching for: Charities. It will then search through the Web to find web pages that contain information based on the key words you typed in. When it has found all the sites, it will display them all in a list. The list may take up hundreds of pages.

1 Open Internet Explorer, if it is not open already.
2 Type www.google.co.uk into the address bar. The Google home page will now load.

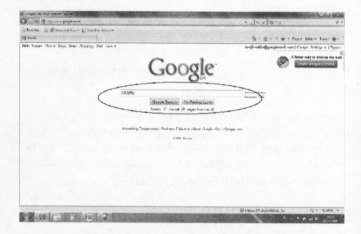

3 Type the word: 'Charity' into the box as shown. Notice that as you start to type the word, Google will predict what it is you are searching for and will show you a list of options. If you see what you want in this list you can click on it. Alternatively, just carry on typing your search words.
4 Click on the button for 'Pages from the UK'. This should mean that you only get websites based in the UK, although others do get through sometimes.
5 Click on the 'Google Search' button. After a few seconds it will show the results pages listing all of the websites that contain information about charities.

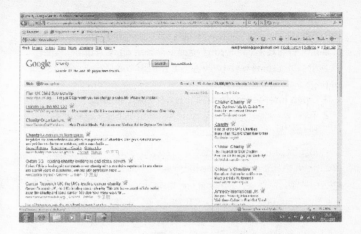

This page shows the first 10 websites that contain information that meet your key words. It also has some sponsored links, to businesses that have paid Google so that their websites will appear on this page. These are the ones in pink at the top and down the right-hand side of the screen.

Each website shown on the results page can be opened by clicking on it. You can read what it says about the website and use this to decide whether it is worth clicking on or not.

In the top right-hand corner you will see how many pages it has found. In this case it has found 24,000,000 pages. This is sometimes called the number of **hits**. It would take years to search through all of these, so we need to narrow down the search.

1 Scroll to the top of the page and type the words: 'British Red Cross' into the box and click on 'Search'. You will find that this reduces the number of hits significantly, and that the website for the British Red Cross is first on the list on the first page of results.

2 We actually wanted to find out how to donate to the British Red Cross so we could do two things here. You could click on the link to the British Red Cross website and follow the hyperlinks to the donation section.

Or you could refine the search still further:

3 Type the words: '"British Red Cross" + donation' exactly as shown.

4 Press ENTER. The results page will now show a direct link to the donations web page of the British Red Cross website.

5.4 How to tell if a website is trustworthy

Just because a website is listed by a search engine does not mean that it contains the information you need, or that the information is correct. Anyone can put information onto a website and there are plenty of strange people out there!

It is not always easy to tell how reliable a website is, but there are some general guidelines you can use:

* Only rely on websites if they are from organizations or businesses that you already know and trust.
* Check for an 'About Us' link to see if you can find out who is responsible for the site.
* Check the name of the site. If it is a .gov site for example, you know that it has come from the government (whether you trust it or not is up to you!). If it is .co.uk it could be from anyone.
* Most sites are trying to sell you something, so you need to be as cynical as you would be if confronted with a pushy salesman in a shop!
* Some websites are what is known as secure sites. There is more information on this in Chapter 7.

5.5 Dead links and redirection

It is quite common to click on a hyperlink and not to get the page you want. This might be for a number of reasons. The web page might no longer exist, or the link might have been set up incorrectly. These are sometimes called **dead links** as they don't take you anywhere. You will most likely get a message on the screen saying that the web page has not been found.

* If this happens, click on 'Back'.

This will take you back to the page that you linked from. It is worth trying again, as sometime you just get a bad connection. If you try again and you get the same message, then the link is probably dead and there is nothing you can do about it.

Sometimes you will be redirected to another website. Sometimes this is for genuine reasons, as the website may have been moved to a different address. Sometimes, it is an advertising ploy to take you to a site that then tries to sell you something. A bit like dead links, all you can do is:

* Click on 'Back' or click on the cross to close the window.

5.6 Common features of websites

As you will soon discover, every website you visit looks different. At first it can be difficult to find your way around some websites. However, many websites have similar features that you should look out for. For example, on many websites:

* The links to other pages are either across the top of the page or down the left or right-hand side.
* A registration process is required, usually requiring your email address and a password.
* You may have to fill in a form to register with the website.
* Where the website has masses of information there will be a **browse** or **search** facility to help you find what you are looking for.
* There may be the opportunity to **download** things, which means that you can take things like music or software from their sites and put them onto your computer.

Throughout the rest of this book you will need to use the skills learned in this chapter. We do use some websites as examples, although we should point out that we are not endorsing these sites and there will always be plenty of others to choose from. We would encourage you to find your own websites either on recommendation from others, or perhaps ones you have seen advertised. Alternatively, you can find suitable sites by typing appropriate search words into a search engine such as Google.

6

keeping organized when using the web

In Chapter 5 we looked at how you could find websites either by typing in the web address, or by using a search engine. What you will probably find is that there are some websites that you want to use over and over again. If this is the case, you want to be able to get back to them quickly. Internet Explorer keeps a record of the web pages that you visit. This is called your history file. This chapter will look at how you can use the history file to revisit websites. It will also show you how you can create a list of favourite websites – the ones that you will visit the most often.

6.1 Revisiting websites using the address bar

This history file stores the address that you type into the **address bar**. This can be very useful if you want to revisit a website as it saves you having to remember the address. After a while the list can get quite long.

To see how it works:

1 Open Internet Explorer, if it is not open already.
2 Type an address into the address bar, e.g. www.bbc.co.uk.
3 After a few seconds, the BBC home page will open.

So far in this book, we have been to the Oxfam website and to the Google website. You might have been to a few other websites too while you have been experimenting. All these websites will be saved in your history file.

4 Click on the small arrow at the far right-hand side of the address bar as shown. A list will appear, showing the website addresses that you have typed in.
5 To go back to any of these sites, all you have to do is move the mouse over it so that it is highlighted, and then click on it.

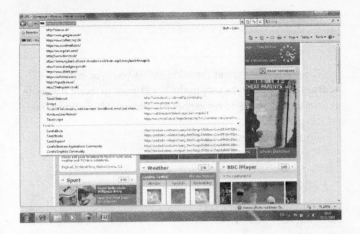

6.2 Saving websites into a favourites list

The problem with using the address bar to get back to websites you have been to before is that you still have to find the one you want in amongst what could be a long list of sites. If you find a website that you know that you will want to use again, you can save it into a special list, called your **favorites**. (Notice the American spelling.) The idea is that you should only put a small(ish) number of websites into your favourites list. These would be the ones that you visit most often.

In this example, we will put the BBC website into the favourites list.

1 Open the BBC website, either by typing www.bbc.co.uk or by clicking on the little arrow and finding it in the list.

2 Click on the icon that contains a small gold star with a green cross on it. This is called 'Add to favorites'.

3 Click on 'Add to favorites'. A new window will now open in the middle of the screen.

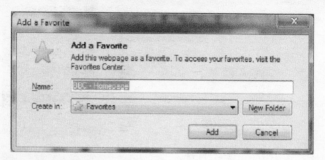

4 Notice that the Name has been put in automatically – you can type over this with a new name if you like.

5 Click on the 'Add' button. The BBC website is now saved in your favourites.

6 You can now go to this website at any time by clicking on 'Favorites'.

7 Click on 'Favorites'. A list of every website that you have saved as a favourite is now displayed. In this example there are a few listed, and you can see the BBC website at the bottom of this list. On your computer, you might only have the BBC website in the list.

8 To go back to the BBC website, highlight it in the list and click on it. You might want to test this out by changing to another website, and then following these steps to get back to the BBC website.

6.3 Changing your home page

As you have seen, when you first click on Internet Explorer, a web page opens up. This is called your home page. The page it opens will depend on how you (or the shop) set the computer up in the first place. You might want to change this to something else.

You usually set your home page to the website that you use most often, like your email website, or perhaps the search engine you use most often.

1 Go to the website that you would like to set as your home page.

2 Click on the little arrow next to the icon of the little house as shown.

3 Select 'Add or Change Home page'.
4 Click on 'Use this page as your only home page'.
5 Click on the 'Yes' button. The next time you open Internet Explorer, it will now open this page automatically.

6.4 Opening several web pages at the same time

Internet Explorer 8 has **tabs** that let you have more than one page open at the same time. This can be quite useful if you want to keep a page open and then go and view a different page. You can have quite a few pages open at the same time if you want to.

To open a new tab:
1 Click on the new tab icon as shown or you can press CTRL + T.

2 A new page will now open.

3 Type the web address of the page that you want to view into the address bar. For example, to open the Oxfam website: type www.oxfam.co.uk. You have now got two pages open: the BBC and Oxfam. You can flick between the two by clicking on the appropriate tab.

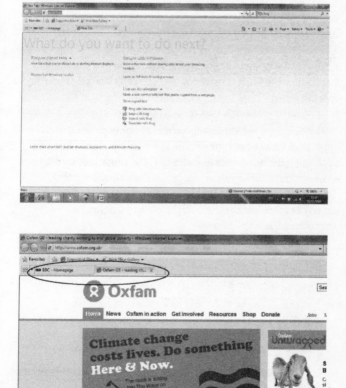

You can close an individual page by clicking on the cross in the tab.

If you close Internet Explorer using the cross in the top right-hand corner, it will close all of your tabs.

6.5 Viewing and deleting websites in your history file

Every website that you visit is automatically saved in a list called your history file. As we have seen, this can be quite useful as you can revisit websites without having to remember the address. However, the history file can get very big. The disadvantage of this is that when you click on the little arrow in the address bar to view the list, it is very long and you can't find what you are looking for.

Another problem is that sometimes you end up on a website that you did not mean to visit, and you might want to delete it from your history file. For example, there are some risks involved with using the Internet, like viruses that can cause problems on your computer, or **spyware** that records what you are doing. Sometimes you get linked to pornographic sites. If you end up on any of these websites, you might want to delete them from the history file.

To view your history file:

1 Click on 'Favorites'
2 Click on 'History'.
3 The history file is organized into days and weeks. In this example, you can see where it says 'Today'. This lists all the sites visited on this computer today.
4 To view it, click on 'Today'. The sites are now listed.
5 To go back to any of these sites, highlight the website and click on it in the list.

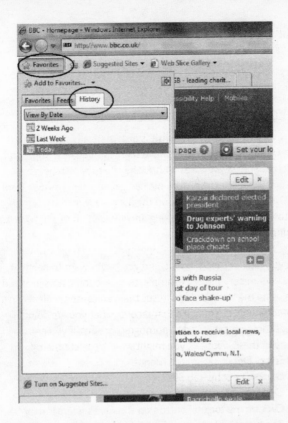

6 To delete any of these sites from the history file, highlight
   the website you want to delete and right click on it to open
   its shortcut menu. In this case, a menu will appear with the
   options to 'Expand' or 'Delete'.

7 Click on 'Delete'. This now removes that website from your
   history file, and it will not appear in the address bar when you
   click on the little arrow.

6.6 Deleting your entire history file

If you find that your address bar has got too full with old addresses, you can clear the list and start again.

Hints and tips

You should use your 'Favorites' to save websites that you definitely want to go back to again. This means that you can safely delete your history file from time to time without losing the addresses of your favourite websites.

1 Select 'Tools' near the top right-hand corner of the screen as shown.

2 Select Options from the menu.
3 On the 'General' tab, which should be open, click on the 'Delete' button as shown.
4 Click on the 'Delete' button on the next screen. It may take a few seconds or even minutes to delete the history file. When it has finished, this screen will disappear.

Hints and tips

Notice the 'Settings' button next to 'Delete' in the Browsing History area. If you click this it will open a panel where you can set the amount of space to be used for storing History files, and the number of days to keep files in it. Setting these to low values will reduce the quantity of History information.

5 Click on 'OK'.

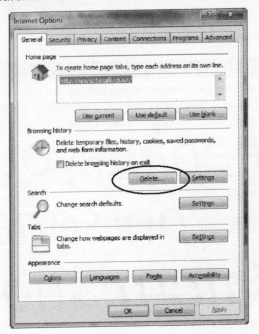

7

keeping your personal information and your computer safe online

The Internet is a global connection of computers with the connections being made by telephone cables and satellites. It all works like the telephone system and logging on to the Internet is a bit like making a telephone call. In fact, your computer has its own number (called an IP address), which is transmitted whenever you are online. In the same way that a telephone call can be intercepted, so can any of the information that is transmitted when you are in the Internet. It is largely unregulated, and pretty much anyone can get access to it. Unfortunately this means that it is open to abuse. This chapter lists the threats that exist and, in each case, explains what you can do about them.

Your computer is also at risk every time you are on the Internet. It is virtually impossible to avoid picking up a few problems whenever you use the Internet. At one end of the scale, you might find that you get a lot of junk email (spam) or pop-ups, which are annoying, and clog up your computer, but do not actually damage it. At the other end of the scale, you could pick up a computer virus, which could destroy information and programs stored on your computer.

7.1 Phishing and identity theft

Identify theft occurs when someone obtains personal information about you which means that they can pretend to be you, usually for fraudulent reasons. They could buy things from the Internet in your name, or perhaps borrow money or even clear out your bank account.

One of the ways that they can obtain the information is by **phishing** – sending an email claiming to be from your bank. They will ask for personal information including your bank account details, or direct you to a fake website that asks you for this information. These emails can look very convincing.

What to do about it:

* Banks will never email you to ask you for personal information such as PIN codes and passwords. If you are asked for it, don't give it.
* Make sure that when you are doing any banking over the Internet that the site is secure. **Secure sites** have https in the address, display a small padlock just to the right-hand side of the address bar and will turn the address bar green:

This is the log-on page for the Alliance and Leicester. Notice how the address has the https at the beginning and there is a padlock icon. Do not disclose any personal information unless the site has both of these showing.

7.2 Hacking

Hacking is when someone gains unauthorized access to your computer. They can do this any time you are connected to the Internet.

You will not even know that it is happening. Hackers do it for various reasons. Often it is just bored teenagers, but some hackers do it with the intention of getting your personal information.

What to do about it:

* Install a **firewall**. This is software and hardware that examines information that is being passed to your computer while you are online. If it finds something it doesn't like, it will block it. If you are using Windows you will already have a firewall. There are lots of different makes of firewall software available to buy. You can find these by searching for them on the Internet.
* Disconnect from the Internet. Only stay online if you need to be. At other times, log yourself off. To do this:

1 Click on the 'Network and Sharing centre' icon on the right-hand side of the Taskbar.

2 This window will show you all your connections including any local area networks that you may have set up. If you are connected using a local area network then you can disconnect by clicking on it here and selecting 'Disconnect'.
3 If it shows that you are connected directly to the Internet, you can click on it and then 'Disconnect'.
4 To log back on, you click on the same icon in the toolbar and 'Connect'.

7.3 Undesirable material

The unregulated nature of the Internet means that you can get access to plenty of undesirable material. Often you will click on a site that you think is perfectly innocent only to find that it contains undesirable content. This may be of particular concern if children have access to your computer.

What to do about it:

* Use your common sense. If you don't like what you see, click on the cross immediately to close the website.
* Install **filtering and blocking software**. This special software allows you to block access to sites that contain undesirable content.
* Set the 'Content Advisor rating' in Windows 7. This is like the filtering/blocking software mentioned before but is already built into Windows.

1 To set the content advisor rating, with Internet Explorer open click on 'Tools' near the top right-hand side of the screen.
2 Select 'Internet Options'.
3 Click on the 'Content' tab as shown.

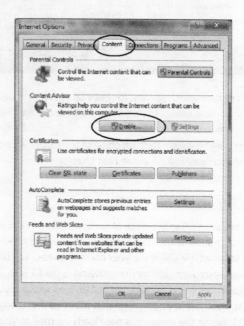

4 Click on 'Enable'. The following screen is displayed:

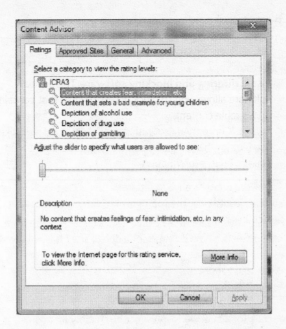

5 You can now select a rating for each category by moving the slider. In most cases, you will want the slider as far over to the left as you can get it.

There is more information on some of the other security features of Internet Explorer in the next chapter.

7.4 Premium diallers

A **premium dialler** is a piece of software that installs itself onto your computer without you realizing it. The next time you log on to the Internet it will not use your normal number to connect, but will dial a premium rate service charging up to £1.50 a minute. This is legal as the software does not install itself without you knowing it – you will see a message that asks you if you want to install it. However, the message is not clearly worded so you may click 'Yes' not really understanding what you are signing up to.

This is a tricky scam as you often see messages popping up when you are on the Internet, and most of them are fairly benign. You are most likely to encounter a premium dialler scam on download sites. These are sites where you can get free software.

What to do about it:

* Read all messages carefully before clicking 'Yes' whenever you are on the Internet.
* Phone your telephone company (e.g. BT) and ask them to block all outgoing calls to premium rate numbers.
* Only use reputable sites.

7.5 Unreliable sites

Many of the problems described in this chapter come from unreliable sites. But how do you spot a dodgy site? It is not always easy as even bad websites can be made to look good.

What to do about it:

* Don't click on a link to a website that comes from an unsolicited email.
* Look for a 'real world' presence, preferably an address.
* Only use sites of well-known businesses or sites that are recommended.
* Avoid sites that offer free downloads, free movies, free music, free games or **file-sharing**.

7.6 Buying online

When you buy anything online there is always a danger that the goods will not be delivered, or what is delivered is not what you ordered.

What to do about it:

* Only buy from trusted websites. This could be the websites of large companies or those that have been recommended by a friend.
* Keep copies of all receipts. All decent online stores will provide a screen where you can print a copy of your order. Most will also send an email to confirm the order.

* Check for a real address so that you can contact them if something goes wrong. It is preferable if they are located in the same country as you!
* Use your common sense. If a deal looks too good to be true – it probably isn't true.
* Ensure the payment area of the site is secure – look for the https and the padlock symbol.
* Have a separate credit/debit card that you use for online transactions and only have a small credit limit on it.
* Use secure payment services such as **PayPal**. These provide insurance against non-delivery.

7.7 Passwords

Passwords usually in combination with a user name are required all over the place. Your computer itself will probably require a user name and password. Email sites, online auctions, chat rooms, etc. all require you to register with a user name and password.

Passwords are very important. There are some rules that you should follow:
* Never give your password to anyone else, ever.
* Change your password regularly and don't use the same password twice. (This one is tricky as it is difficult to remember them all.)
* Don't choose something obvious like names, dates of birth, etc. Use combinations of letters and numbers, as they are harder to guess.
* Don't write passwords down anywhere.

Don't have nightmares...

Internet crime is increasing and you are never immune to threats even if you take all of the precautions listed in this chapter. However, if you take precautions, the chances of becoming a victim are very small. Remember that millions of people now use the Internet regularly with no problems.

7.8 Junk email

Junk email, also known as **spam**, is almost impossible to avoid. Junk emails appear in the inbox of your email software. You didn't ask for it and you don't know the person or organization that sent it. Most of the time, they are just advertising things, particularly Viagra, distinctly dodgy investments and pornographic material.

What do to about it:

* Most email software will filter out spam emails for you and put them into a special folder called Spam. However, some will still get through.
* It is possible to minimize the amount of junk mail you get by blocking or reporting it to your service provider. This means that you should not receive any more spam from these organizations again. For example, in Gmail you can click on the 'Report spam' button when you get it:
* If you do not know who an email is from, delete it without opening it.
* Don't reply to email from spammers.

The real danger of spam is that email is a very common way to pick up a computer virus, and these can be quite serious, as we will see later on.

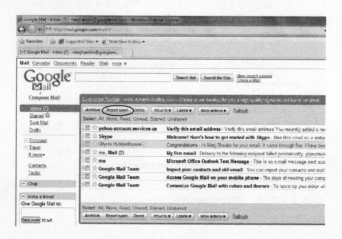

7.9 Pop-ups

Pop-ups are windows that just pop up (appear) when you are using the Internet. Like spam, they are not something that you asked for – they just appear. Most of these pop-ups are just advertising something. Some of them look quite enticing because they tell you that you have won something. This is usually just a trick to get you to visit their website.

Pop-ups are not necessarily bad, but they can be annoying if you get a lot of them. The simple solution to a pop-up is simply to click on the cross and get rid of it.

What to do about it:

* Read the pop-up, as it might be a genuine offer from a reputable company. If not, click on the cross.

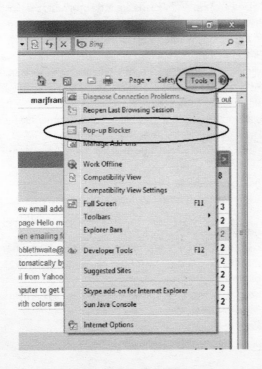

* You can buy pop-up blocking software, which reduces the
  amount of pop-ups you will get.
* You can set up Internet Explorer so that it stops most pop-ups:

1 Open Internet Explorer.
2 Select 'Tools' near the top right-hand corner of the screen.
3 Select 'Pop-up Blocker'.
4 Select 'Turn on Pop-up Blocker'.
5 Click on 'OK'. This should now stop most pop-ups from
  appearing and will warn you when it has done so.

Some websites will not work properly if pop-ups are blocked.
If this happens, you can choose to allow pop-ups from the website.

7.10 Viruses, trojans and worms

Viruses, **trojans** and **worms** are little programs that install
themselves on your computer without you knowing about it.
This normally happens when you download something from
the Internet or when you open an email. Like human viruses, a
computer virus will infect your computer causing all sorts of
problems. Some are worse than others. Really bad ones will delete
everything on your computer.

Trojans are viruses that are hidden inside another file (as in
Trojan Horse). Worms are viruses that are designed to infect lots
of different files in your computer, making them difficult to get
rid of.

What to do about it:
* Only download from reputable sites or secure sites.
* Do not open emails (especially email attachments) if you
  do not know who they are from.
* Use anti-virus software. You can get free software from the
  Internet or you can buy it from companies such as McAfee
  or Norton.
* Keep your anti-virus software up-to-date as new viruses
  come out every day.
* Keep your version of Windows up-to-date as many updates
  contain fixes for well-known viruses.

7.11 Spyware and adware

This is software that installs itself on your computer without you knowing about it. It can do this any time you are on the Internet. The software collects personal information that you fill in when online, and tracks which websites you visit. The information it gathers is usually used for marketing purposes.

What to do about it:

* You can download free software from the Internet that will check your computer for spyware/adware, or you can buy software that will do it for you. This type of software is called **Spyware** or **adware** removal software.

* Keep your version of Windows up-to-date. Once you have bought Windows you are entitled to free updates from their website (www.microsoft.com).

7.12 Creating a backup of your work

Most of these threats are relatively minor. The most dangerous of all these threats is that you get a virus. Viruses vary in seriousness. For example, some viruses just do annoying things like automatically redirect you from one website to another, or maybe it will close down Internet Explorer automatically without warning.

In serious cases, viruses can destroy any of the information stored in your computer.

What to do about it:

* Keep the original copies of all CDs/DVDs that came with any software that you bought, in a safe place.

Make a **backup** of anything that you have saved. The best way to do this is to copy it onto a CD or DVD. This will only work if your computer has a drive that can write CDs or DVDs.

DVDs are more expensive to buy but can store much more data, so use DVDs if you can.

1 Open the CD/DVD drawer on your computer by pressing the button.

2 Insert either a CD-R, CD-RW, DVD-R or DVD-RW shiny side down.

3 After a few seconds you may be asked what you want to do with the CD/DVD. If so, select the option to 'Burn files to disk'. If not, don't worry as you can open it shortly.

4 Click on 'Start'.

5 Click on 'Documents library'. This is where most of your files will be saved to. Notice that there are also libraries for music, photographs and videos. You can create a copy of all the information in the Documents library on a CD/DVD.

6 Press CTRL and A at the same time. You will see that all the files and folders are highlighted.

7 Point the mouse at any one of the files, right click and select 'Copy'.

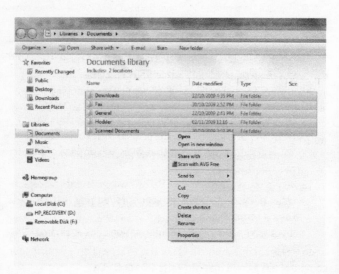

8 On the left-hand side, click on the CD or DVD. In this case it is a DVD.

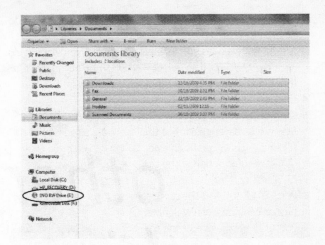

9 You will be prompted to give your CD or DVD a name. Type in a suitable title, e.g. 'Documents backup April 2011'.

10 It will take a few seconds to format (prepare) the CD or DVD. A blank window is now displayed. This represents the contents of the CD or DVD, which is currently empty.

11 Right click anywhere in the white space.

12 Select 'Paste' from the list. The computer will now start copying all the files and folders in your Documents library ready to put them on the CD or DVD. This could take a while. You will get a progress bar that gives you an idea of how long this will take.

13 When it has finished, the CD/DVD will automatically eject. Label it – you can buy special pens for writing on CDs – and keep it somewhere safe.

If you ever need to get the information back again from this CD/DVD:

1 Put the disc in the drive.

2 Click on 'Start'.

3 Click on 'My Computer'.

4 Select the CD/DVD and copy files back from there to the folders that you want them in.

other things you can do on the Internet

Once you have connected to the Internet, there are many things you can use it for. The Internet connects your computer to a whole collection of other computers and is a great way of communicating with the world above and beyond email. You can make phone calls over the Internet, send instant messages, comment on any news, articles or information you find, join communities of like-minded people, share your opinion or comment on other people's. It is a great way to share your knowledge as well as learn from other people's experiences.

The Internet can also help you with more practical things. You can book your holiday online, monitor your finances and pay your bills, buy or sell almost anything and have your shopping delivered to your door. It can also provide you with employment or educational information. This chapter will show you how to do all of this and more.

8.1 Making phone calls over the Internet

VOIP stands for Voice Over Internet Protocol and describes the facility to make phone called using the Internet. One of the most popular providers of a VOIP service is Skype. You can use VOIP to communicate using text, voice and video. You need broadband access, special software, and either an Internet phone or a microphone and headset or speakers. The first thing you need to do is download and set up the free Skype software. You have to set yourself up with a Skype name, which has to be unique. Skype has an address book feature called 'Contacts' and a directory where you can search for someone if you don't have their Skype name. You can send text messages. The person may respond instantly. If not, the messages are there for them next time they go to their Skype account. You can make telephone calls in the same way as a normal call. You can use an Internet phone for these, or a headset. You can also have a video call if you have a webcam. You will be able to see and hear the person you are talking to.

8.2 Chatting over the Internet

Sometimes you want to be able to have a live conversation with someone where you get an instant response. There are two main ways of doing this:

* Use **instant messaging (IM) software**. This enables you to chat with people who you know – they would need to have the same IM software on their computer. You can get access to IM software in a number of ways. It is available within Skype or you can download Messenger, which is a Microsoft product. It is also available on the websites of most of the social networking sites such as FaceBook.

* Use a **chat room**. Generally speaking you don't need any special software to use a chat room, you just log on to a chat room website and you can join in the live conversations that are taking place.

Chat rooms

Chat rooms are public websites where you can have **real-time** conversations with people. Some chat rooms are general, where people go in and chat about anything they like. Other chat rooms are themed, which means that the chat is about a specific topic. Some chat rooms are regulated, which means that there is a moderator who keeps an eye on what is being said, in case it gets offensive. Others are not supervised, and anything goes.

There are thousands of chat rooms on the Internet covering every possible theme. You can find a chat room by using a search engine and typing in suitable search words, e.g. 'chat room + gardening'. Also, many well-known websites have a chat room facility, e.g. Google and Yahoo both operate chat rooms on a range of themes. Also, many charities now have chat rooms where you can find people with the same interests and concerns. Many medical and health charities for example, have chat rooms for people affected by illness.

Many chat rooms are of an adult nature although this is usually made obvious before you enter it. These tend to be unregulated so if you are easily offended, steer clear.

8.3 Getting involved with online communities

As well as using the Internet to find information, more and more people are now adding their own contributions. Whole communities have developed on the Internet.

Social networking sites like MySpace, Facebook, Twitter and Bebo have become very popular. This is where people create an online profile of themselves for everyone to view. Finally, there is the phenomenon known as a **wiki**, an example being Wikipedia. This is an encyclopaedia that is written and constantly updated by anyone on the Internet who feels that they know enough to contribute to it.

Social networking websites

Social networking websites are those where people create profiles of themselves for the world to see. The profile can include personal information, photographs, links to other web pages, comments from viewers, lists of likes and dislikes and information about relationships. As well as viewing other people's profiles, you can also create your own. These sites are very popular among young people, but older people are also getting involved. Two of the best known sites are MySpace and Facebook. Twitter, which is very popular at the moment, is a cross between a social networking site and a blog, where people write brief messages about what they are doing right now.

Many people use social networking sites as a way of making contact with new people with a view to meeting up in the 'real world' as well as the 'virtual world'.

8.4 Reading and writing blogs

A **blog** (short for web log) is a web page written by an individual or group of people, usually in the style of a journal that contains regular entries (like a diary). Each entry is called a **post**. Posts are dated and shown in reverse chronological order. The information in a blog could be anything from the day-to-day life of the **blogger**, through to information on particular topics and themes. Many blogs are text only, though they can contain images, movies and sounds.

Some blogs have become extremely popular with millions of people reading them on a regular basis. Twitter is worth a special mention at this stage as it is so popular. The idea of this is that you write short messages telling people what you are doing and thinking right now. It is sometimes called **micro-blogging** as you are not supposed to write much.

Finding blogs to read

In common with anything else on the Internet, your start point could be to use a search engine to find blogs on themes that you are interested in. For example, typing 'blog + classical music' into a search engine will give you thousands of potential blogs to read.

Many of your favourite websites may also carry blogs. For example, many leading newspapers and TV stations have blogs.

Another way of finding a blog is to use a specialist bloggers site. One of the most popular is Blogger, which is part of Google's network of websites.

Creating your own blog

The easiest way to create your own blog is to register with a blogger site, many of which are free. The advantage of using a site like Blogger is that you can create a professional looking web page without having to learn how to use any specialist software.

Blogging websites make it easy for you to set up your blog by offering you predefined layouts for the page. Your blogs can consist only of text or as you get more confident, you can start adding other features to your websites such as photographs and video. Blogs allow you to set up your personal profile. This can be shared with your readers so that they can get to know more about you. Blogs can be used as a way of making new friends and finding like-minded people.

8.5 Arranging and booking your travels

Once you have got to grips with using a search engine (such as Google), you can use it to search for absolutely everything, including travel and holidays. The problem is if you just type 'holidays' into a search engine you will end up with millions of **hits** and it will be impossible to look through all of them.

One solution to this problem is to put more specific words into the search engine so that it only finds websites that are relevant to your requirements. Another solution to this problem is to find the web address of particular companies that you could use to arrange your travels. For example, Saga arrange holidays specifically for the over 50s, so you could do a search specifically for this company.

Choosing and booking a ticket/holiday

Once you have found the website you want to use, you then need to find the holiday or flight that you want, and book it.

This process will be different depending on which company you use, as all of their websites are slightly different. However, the principle is the same in each case. You choose the holiday you want and then you have to fill in a form and make payment using your credit card.

Every company's web pages are different so you need to take a bit of time to look around the site and find what you want.

Finding and booking train tickets

There are two main websites that you can use to find and book train tickets:

* www.nationalrail.co.uk
* www.thetrainline.com

Both work in a similar way, where you type in the route that you want to take, and it will show you what trains are available. You can then book them and pay for them online.

Using a route-planner for a road journey

If you are planning a road trip, you can get detailed instructions and maps free on the Internet. All you have to do is type in the start point and your destination. A few websites offer this service. Two of the main ones are provided by motoring organizations:

* www.theaa.com
* www.rac.co.uk

Both have the route planner facility on their home page, which makes them very easy to find.

Some websites that offer route-planning also have up-to-date traffic information to help with your planning.

8.6 Banking online and other financial services

All of the main high street banks now offer an online banking service. It is possible to carry out most banking transactions online without having to visit your branch or use the telephone, including:

* Checking your balance
* Viewing statements

* Paying bills
* Changing or cancelling standing orders and direct debits
* Transferring money

You can also access a range of other financial services including arranging loans, mortgages, life insurance, ISAs and investments. You might want to do this through your existing bank, or you can use the Internet to shop around.

Getting started with online banking

Before you start online banking, you have to register your details with the bank for security purposes. Depending on the bank, you may need to do this over the phone or in the branch. Some banks allow you to do it using the Internet. If you want to do online banking with your existing bank, you will find details of how to register on their website.

Part of the registration process is to choose a security code (like a PIN) and answers to security questions (for example, mother's maiden name or first school). Once registered you will have to type in your sort code, account number and then you will be asked for the PIN and for the answer to the security question.

Once registered, you can **log on** to your account.

Carrying out transactions

As well as viewing your statements, you can pay bills, set up, change or cancel standing orders and direct debits and transfer money to other accounts. For example, to change or cancel a standing order:

1 Find the link to 'Standing Orders'. All of your existing standing orders will now be displayed.
2 To change or cancel a standing order, double click on the name of the person or organization who is receiving the money from you. The individual details of the standing order are now shown and you can amend or cancel it by clicking on the appropriate link.

Once you are familiar with one transaction, all of the others are very similar. For example, managing your direct debits and bill payments is very similar to sorting out your standing orders.

When you have finished banking, it is important that you log off from the website, rather than just clicking on the cross. There will be a link to 'log off' or it might be called 'sign out'. This guarantees that the connection between your computer and the bank's is closed, which makes you safe from hackers.

Other financial services

In addition to the banks, there are lots of other financial services available over the Internet including loans, insurance, investments, mortgages and pensions.

Many financial services businesses offer cheaper deals if you buy online, because it saves them money.

Choosing which of these businesses to use involves the same decision-making process that you would use if you bought from them without doing it online. For example, you might choose to use a company because they are the cheapest, or because they offer the best service, or because you trust them.

8.7 Buying from an online auction

eBay is currently the world's biggest online auction site boasting more than 230 million users worldwide with millions of items for sale at any one time.

It works much in the same way as a traditional auction in that items are offered for sale, you look at them and read the description, and then decide whether you want to bid on the item or not. Many other people will be doing the same thing and bidding against you. The big difference with a real auction is the period of time over which the bidding takes place, as it can be several days. At the end of this period, if your bid is the highest, you win. You then pay for the product and the seller sends it to you, or you go and pick it up.

Getting started

First, you need to register. To do this, you must already have an email address.

Decide how much you want to spend and stick to it. In common with traditional auctions, it can be tempting to keep upping the price that you are willing to pay, only to regret it later. Many bidders leave it until the last minute to make their bids; so don't get caught up in a bidding war.

Make sure you are aware of all the charges that will be added. Keep an eye on postage costs and VAT. Many businesses now use eBay as their main way of selling products, and they will have to add on VAT. Also, make sure that the item is for sale in your country, or you may have to pay additional shipping costs and tax.

Finally, eBay has become an international phenomenon and as a result, there is a lot of information written about it. A good starting point is the company's own Help centre, which can be accessed by clicking the 'Help' tab on their home page.

You can either browse by category to find what you want, or use the search facility. There is a lot of information available about the item you want to buy and the person or online business who is selling it. You can make a bid on an item and then continue to bid as the price changes. Alternatively, you can place a maximum bid. Some items are available to buy straight away or shown as a classified ads, which means you do not have to bid as such, just make an offer. There is a 'watch this item' feature which allows you to keep an eye on particular items whether you are bidding on them or not. If you win the auction you will be notified and you then have to make payment, which may be via PayPal or more traditional methods. As with any purchase, you need to satisfy yourself that you are happy with the item and the person selling it before you start bidding.

8.8 Doing your grocery shopping

The first stage of the process is to check that the retailer you want to use does deliver to your area. For example, Tesco, Sainsbury's, Asda and Waitrose all offer delivery services, but you do have to check that they will deliver in your locality. You also need to consider the cost of delivery, which is usually around £5. You can then order your products by filling up a **virtual** basket or trolley.

You book your delivery slot, pay online and then wait for the van. Generally speaking you need to book a day or two in advance and then wait in for a delivery slot, which is normally given within a two-hour time period, e.g. 9 am–11 am.

The first time you shop online it might take quite a while, but it will be quicker when you do it subsequent times, because the websites use the 'favourites' idea to remember the things that you bought last time.

Ask around among family and friends to find out who delivers in your area and whether they are any good. You have to register with the website, which will involve giving them your bank account details so that you can complete online transactions. There are special offers on the website like you would find in store and you can also redeem vouchers. When the goods are delivered, they will offer substitutes for anything that was not available in the store. You can reject those that are not satisfactory.

8.9 Working and learning

There are now over 20 million over 50s in the UK, many of whom are taking part in either employment or education of some sort. The Internet is a good way of keeping up-to-date with the changing world of employment and education. The website www.direct.gov.uk contains masses of information relating to government services including advice on state pensions. There are thousands of websites dedicated to people who are looking for work and these can be found using a search engine. Local councils and job centres also have websites advertising jobs. You can register with jobs websites and they will email you updates. You can also upload you CV to their sites for potential employers to see. If you are thinking of starting your own business there is a lot of information and advice available. You can use the Internet to find out about local, national and international volunteering opportunities. You can use the Internet to find out about courses and night classes being offered around the country. You can use the Internet to take an online course, a bit like a traditional 'correspondence course' but all done via websites and email.

8.10 Accessing TV, radio and games

Multimedia (the mixture of text, sounds and images) is what the Internet is all about. The Internet lets us access multimedia content (music, videos, games, etc.) whenever we want it. There are two main ways of doing this. The first is going onto websites where we can watch TV, listen to radio and play games. The second is downloading music and video to your computer so that you can view and listen to it without having to be online.

There are hundreds of online radio stations, including ones that you can also listen to on a traditional radio. Online radio stations are available covering all possible genres including different types of music and talk radio. You can listen to radio stations live or use a 'listen-again' feature to hear programmes you have missed. Radio stations are usually played through a media player which is a piece of software designed to play multimedia files. Using the media player you have some control over a programme. For example you can pause it and you may be able to rewind and fast forward. There are loads of online TV stations covering a range of genres, many of which are available online. Terrestrial TV stations do not normally broadcast their programmes online while they are going out live, but they do have a 'watch again' feature. There are thousands of websites dedicated to playing games online. Some games websites are traditional computer games and are free. Others are where you play against other online players, sometimes for real money.

8.11 Accessing music and films

The Internet has roughly 1 billion users throughout the world. What this means is that however obscure your musical or film tastes, you will be able to find it somewhere on the Internet.

Another advantage is that you can get access to the music and videos that you want almost instantly as you can download it straight onto your computer. You don't have to go to the shops and you don't have to wait for delivery.

A disadvantage is that a lot of the music and films that can be downloaded are on websites which are not legally entitled to offer them, because they do not have copyright permissions. The difficulty is knowing which sites are legal and which are illegal. If you are downloading a feature film for free, and it's not some kind of special offer, then it is probably illegal. It's not so clear with music, where many artists make their work freely available online, hoping that this will encourage fans to buy their CDs and come to their shows.

Music can be downloaded and played from your computer or you can transfer it to a portable music player. Most legal websites will charge you to download music, although some music is free. You pay for this in the same way as any other online purchase. A podcast is a self-contained download that is part of a series. You can subscribe to them or download the episodes you want to hear. Many radio stations and TV programmes have their own podcasts, which are condensed versions of the shows. You can find podcasts on a range of topics. For example there are podcasts for learning languages, comedy programmes and political commentary. There are hundreds of online retailers of films and music. Often you can buy CDs and DVDs online cheaper than you can in high street stores. There are DVD rental businesses such as Lovefilm operating online. These will send DVDs by post.

8.12 Dating online

There are literally millions of people using the Internet for dating purposes. One of the biggest sites in the world is www.match.com which boasts 6.5 million members in the UK alone. That's a lot of fish in the sea! There are lots of dating agencies to choose from and some of them specialize in particular interest groups or age groups. A quick search for 'over 50s + dating agency' will provide you with plenty of possibilities.

They all work in a similar way:
* You register your details with the website and create a personal profile

* The website stores the profiles of all the people who are registered
* You search for the type of person you are looking for
* The website shows you the profiles of everyone who is a possible match
* You make contact (usually via email) with individual people
* You establish a rapport using email, or maybe in chat rooms
* You meet, you fall in love and live happily ever after.

Obviously there is some risk attached to this as you are meeting people that you don't know and you need to exercise some caution. Most online dating services take personal security very seriously.

The main tips for safety are:

When **online**:
* Stay anonymous until you feel confident enough to reveal your identity
* Use sensible email names and user names, i.e. nothing provocative
* Be honest in your personal profile and use a recent picture
* Report anyone who is abusing the system

When **offline**:
* Find out as much about them as possible before you meet
* Arrange to meet in a public place
* Tell a friend where you are going
* Make your own travel arrangements
* Watch your alcohol intake
* Don't be pressured into anything.

On the whole dating online is a fun and safe activity. Basically you just need to use your common sense as you would in any other aspect of your life. The best advice is to trust your instinct.

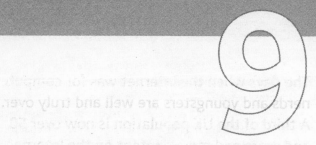

websites
for the
over 50s

The days when the Internet was for computer nerds and youngsters are well and truly over. A third of the UK population is now over 50 and more and more content on the Internet is now targeted at this age group.

The idea of this chapter is NOT to give step-by-step guidance on how to access all of these websites but to list sites that might be of interest. Some of these sites have been designed specifically for the over 50s and some of them are designed for everyone to use, but will be of particular interest to the over 50s.

By now you will probably be quite used to searching the Internet, opening pages, following links and switching between pages. This chapter will be a good chance for you to put these skills to the test!

9.1 Portal websites

A **portal** is a particular kind of website that acts as a starting point for someone who is looking for information on the Internet. Portal sites usually have a particular theme. For example, there are many websites that call themselves 'portal sites for the over 50s'. This means that their website contains lots of information and lots of links to other websites that are of interest to the over 50s. If you can find a good portal site it can save you hours of Internet searching, because the portal website has already done the searching for you and listed useful sites in one place.

An example of a portal site that we have already used in this book is www.directgov.uk, which is created by the UK government. As well as providing lots of information on public services, it also acts as a link to many other public services websites. On its home page, you can see all of the links. Some of these links are to other pages on this site, but many of them will be links to pages on other websites.

You can find over 50s portal sites by typing 'over 50s + portal' into a search engine. Here is a typical example from www.myprime.co.uk, which contains information and links specifically for the over 50s.

Follow the link to Directory, which in turn contains links to hundreds of other sites. These are categorized by topic, or you can click on their 'hot links', which will show you some of the most popular links that other people have used.

Other portal sites include:

* www.50connect.co.uk
* www.mabels.org.uk
* www.retirement-matters.co.uk
* www.seniority.co.uk

9.2 Links

You will find that websites often tend to have links to other related websites, even if they are not portal sites. For example, www.overfifitiesfriends.co.uk is an online community site where people can post profiles of themselves, join discussion forums and meet up locally in groups to take part in social activities. It also contains lots of links to other websites aimed at the over 50s.

These links are to all sorts of sites including dating websites, other online community sites, and sites offering products and services for older people.

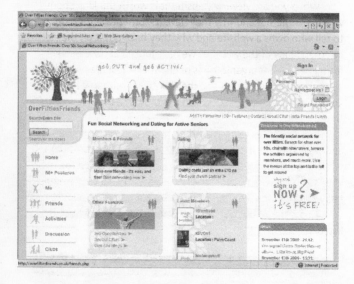

Once you find one good site, you will often find links to other good sites.

It is easy to lose track of which website you are in. Sometimes when you follow a link, it will open the web page in a new window. To get back to where you started you will need to click on the cross to close the new website. Sometimes the link will move you onto the new web page. You can use the Back button in Internet Explorer to move you back to where you started.

9.3 Choosing websites

There are millions of websites out there and new ones coming along all the time. Finding decent websites can be difficult. You can use a **search engine** to find sites, or you can ask family and friends what sites they use. Many of the national newspapers often print lists of useful websites, and it is possible to buy web directories, a bit like Yellow Pages (although these can go out-of-date quickly). You can also get website directories on the Internet.

When you are viewing websites it is worth thinking about who owns the site and why they have it. Sometimes it's hard to tell who owns a website and why they have put it on the Internet. Many of the websites we have looked at have been from government organizations or charities. Most are owned by commercial businesses and their motive is primarily to get you to buy something, so it doesn't hurt to be a bit sceptical at times.

Most websites tell you on their home page who they are and what they do. If it is not clear, try and find a link to 'About Us', which most sites have. There may be a main link to it or you may have to go to the bottom of the home page. The 'About Us' should tell you who owns the site and what kind of organization it is.

This is the 'About Us' page for www.fiftyon.co.uk which is listed in Section 9.4.

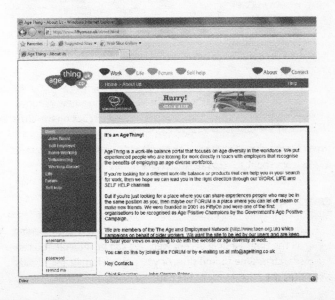

This makes it quite clear who is responsible for the site, and why they are running it.

9.4 Websites of interest to the over 50s

This section lists a small selection of websites that you might find useful, with a brief explanation of who owns the website and what it has on it. We have tried to focus on general sites that in turn will contain links to more specialized interests.

http://silversurfers.digitalunite.com/
Information relating to Silver Surfers week, which is run by Age Concern.

www.ageconcern.co.uk
The main website of the charity. Contains useful information about visual and audio aids when using computers.

www.wiseowls.co.uk
Information, advice and campaigns for the over 50s.

www.nhsdirect.nhs.uk/
Access to the services of the National Health Service.

www.saga.co.uk/
Company specializing in products and services for the over 50s.

www.dlf.org.uk/
Website of the disabled living foundation.

www.bbc.co.uk/health/health_over_50/index.shtml
Pages from the BBC website specifically about health for the over 50s.

www.fiftyon.co.uk
Website primarily concerned with employment issues for the over 50s.

www.lifes4living.co.uk/
Online community website for the over 50s.

www.seniorconcessions.co.uk
Website that searches for discounts available for older people on a
   range of products and services.

www.theoldie.co.uk
Website of the *Oldie* magazine.

www.begrand.net
Website specifically for grandparents.